AF696580

ELECTRICAL LEVEL TWO WORKBOOK

NATIONAL CENTER FOR CONSTRUCTION EDUCATION AND RESEARCH

Contents:

Note: The designations "National Electrical Code®," "NE Code®," and "NEC®," where used in this document, refer to the *National Electrical Code®*, which is a registered trademark of the National Fire Protection Association, Quincy, MA. *All National Electrical Code® (NEC®) references in this module refer to the 2005 edition of the NEC®.*

Name: ______________________________ Date: ______________

Social Security Number: ______________________________

1. In a circuit containing resistance only, the voltage and current are ______________ with one another.

2. The power consumed by ______________ is called *true power*.

3. The power factor is the ratio of ______________ power to ______________ power.

4. LC circuits consist of a(n) ______________ and a(n) ______________ connected in series or in parallel with a voltage source.

5. As in all series circuits, the ______________ in a series LC circuit is the same at all points.

6. On a transformer with 100 volts applied to the primary, 200 turns in the primary, and a secondary with 100 turns, the secondary voltage is ______________ volts.

7. A(n) ______________ can be used to electrically isolate a piece of equipment from the power distribution system.

8. Reactive power is the portion of apparent power that is caused by ______________ and ______________ in the circuit.

9. The primary winding in a transformer is connected to the ______________.

10. Apparent power is the product of the source ______________ and the ______________.

11. The secondary winding in a transformer is connected to the ______________.

12. True power is measured in ______________.

13. VAR is the unit of measurement for ______________ power.

14. The average value is calculated from all the values in a sine wave for only half a cycle. The average value formula is AVG = ____________________ × peak value.

15. In purely resistive circuits, the current and voltage are related by Ohm's law, which is E = IR. Define the variables.

    E = ____________________
    I = ____________________
    R = ____________________

Name: ______________________________ **Date:** ______________

Social Security Number: ______________________________

1. The field coils of a series motor are connected in series with the ______________.

2. A(n) ______________ is used to provide the switching action required for continuous DC motor rotation.

3. Gap measurements should be made on the ______________ end of the motor.

4. The most common class of insulation used in motor manufacturing and rewinding is NEMA ______________.

5. All motors with a voltage rating greater than ______________ must have a thermal protector.

6. What type of motor requires brushes, slip rings, and an amortisseur winding? ______________

7. The two basic types of motor connections are the ______________ and ______________.

8. Explain slip and why it is present. ______________________________

9. Most industrial motors are NEMA Design _____.
   a. A
   b. B
   c. D
   d. F

10. What type of rotor is most commonly used in an induction motor? ____________________

____

11. A megger test determines the _____.
    a. hp rating of a motor
    b. FLA rating of a motor
    c. insulation resistance of a motor
    d. field strength of a motor

12. When a single motor is supplied from a branch circuit, the ampacity of the branch circuit must be ____________________ of the motor full-load current as determined by ***NEC Section 430.22***.

13. A consequent-pole motor has ____________________ speeds.

14. How is the speed normally changed in a multiple-speed induction motor?

____

15. Name three conditions that can cause a synchronous motor to become out of step.

____

____

____

Name: ______________________________ **Date:** ______________

Social Security Number: ______________________________

1. A short circuit is a conducting connection, whether intentional or accidental, between any of the ____________________ of an electrical system.

2. True or False? The neutral or grounded conductor is a system or circuit conductor that is intentionally grounded.

3. A conductor used to connect equipment or the grounded circuit of a wiring system to a grounding electrode is called a(n) ____________________ conductor.

4. The two types of grounding systems are system grounding and ____________________.

5. If the loads were only 240V on a single-phase 240V system with a center-tapped transformer, the neutral conductor would carry ____________________ amperes.

6. Systems less than 50V are required to be grounded if the supply voltage to the transformer exceeds ____________________ to ground.

7. Circuits that cannot be grounded include certain circuits for cranes, electrolytic cells, secondary circuits on lighting systems, and isolated power systems in ____________________.

8. Systems are solidly grounded to limit the ____________________ during normal operation and to prevent excessive voltages duc to lightning and line surges.

9. Grounding methods include an underground water pipe in direct contact with the earth for no less than ____________________.

10. The size of the grounding electrode conductor is based on the size of the ____________________ ____________________.

11. The size of the EGC is determined by the ____________________.

12. All receptacles used in residential applications must be of the ____________________ type.

13. Electrical ____________________ is the key to successful ground fault circuits.

14. The size of the bonding jumper is based on the size of the ____________________ in each raceway.

15. The two main purposes of the neutral in a grounded system are to permit utilization of power at line-to-neutral voltage and ______________________________

____

____.

Name: ______________________________ **Date:** ______________

Social Security Number: ______________________________

1. In what chapter of the NEC® will you find the installation requirements for conduit?
   a. Two
   b. Three
   c. Four
   d. Five

2. The maximum number of degrees allowed between pull points in a conduit run is _________.

3. Offsets and saddles _____.
   a. do not need to be counted if less than 15°
   b. add just as much resistance to pulling conductors as any 90° elbow
   c. add no resistance
   d. all count as 30°

4. A conduit run with _____ is acceptable to pull wire.
   a. two 90° bends, four 45° bends, and one box offset
   b. two 90° bends, three 45° bends, and one saddle
   c. two 90° bends, two 45° bends, and two 30° bends
   d. four 90° bends and two box offsets

5. The maximum size rigid conduit that can be bent using a portable mechanical bender is _____.

6. The maximum size IMC that can be bent using a portable mechanical conduit bender is _____.

7. All of the following may be bent using a mechanical bender *except* _____.
   a. EMT
   b. IMC
   c. PVC
   d. Rigid conduit

8. The maximum size conduit that can be bent safely is _____.

9. When making segment bends, _____ must be determined.
   a. the radius of the bend
   b. the total number of degrees in the bend
   c. the developed length
   d. All of the above.

10. When making a saddle bend using 4" conduit with a minimum length (E-C) of 20" and bend spacing (C-C) of 15½", the angle of the third bend will be _____.
   a. 36°
   b. 20°
   c. 40°
   d. 30°

11. The code requires a minimum bending radius of six times the trade size of the diameter of rigid conduit because ____

____

12. A bend used for a change in direction of less than 90° is called a(n) _____.
   a. kick
   b. saddle
   c. stub-up
   d. offset

13. A mechanical bender may be used to make bends in _____.
   a. 2" EMT
   b. 4" rigid conduit
   c. 1½" PVC
   d. IMC conduit only

14. To make matching bends in both ½" and 1" IMC when using a mechanical bender, _____.
   a. adjust the deduct length for the ½" conduit
   b. use the ½" shoe for the ½" conduit and the 1" shoe for the 1" conduit
   c. use the 1" shoe for both conduits
   d. get matched pairs of conduit shoes from the manufacturer

15. Speed benders can bend conduit up to ________________.

16. True or False? The one-shot shoes used with hydraulic benders can make 90° bends in a single operation.

17. One method of bending segment is to use _____.
    a. an elapsed time method to keep track of each bend
    b. a stress meter to monitor each bend
    c. a level to monitor each bend
    d. the amount of travel method to monitor each bend

18. To make a 20" offset using 45° angles, a multiplier of 1.414 is used. If the first bend is marked at 12" from the end of the conduit, the second bend would be marked at _____ from the end.
    a. 16.96"
    b. 28.28"
    c. 40.28"
    d. 32"

19. Airtight plugs are installed in the ends of large sizes of PVC conduit during bending to _____.
    a. prevent the conduit from collapsing when heated
    b. hold the heat inside the conduit to keep it flexible
    c. prevent condensation within the conduit
    d. prevent deadly PVC gases from escaping

20. One of the disadvantages of aluminum conduit is that _____.
    a. it cannot be bent with one-shot benders
    b. it is difficult to bend consistently and accurately
    c. special bending shoes must be used
    d. aluminum kinks when bent

notes

Name: ______________________________ **Date:** ______________

Social Security Number: ______________________________

1. A Type T conduit body is used to provide a junction point for ______________ intersecting conduits.

2. All boxes must be sized, installed, and supported to meet the current ______________ requirements.

3. Class 1 locations contain flammable ______________ or ______________.

4. Blank covers are used when boxes serve as ______________.

5. Conduit bodies, also called ______________, provide access through a removable cover or covers to the interior of the system.

6. Each conductor originating outside of the box and terminating inside of the box counts as ______________ conductor(s).

7. When sizing outlet boxes, deduct ______________ conductor(s) for each switch.

8. To distinguish between Type LL and Type LR conduit bodies, hold the fitting as if it were a(n) ______________ and see where the cover is situated.

9. Name three Class I locations.

   ____

   ____

   ____

10. For a straight pull, the minimum length of the box must be at least ______________ times the trade diameter of the largest raceway.

11. The minimum length for a junction box in which the conductors are pulled at an angle and that contains two 3" conduits and one 4" conduit entering each side is ______________.

12. Octagon and round boxes are used mostly for ______________________________

____.

13. When sizing outlet boxes, ____________________ conductor(s) must be deducted for each strap-mounted device.

14. An extension ring is used to increase the ____________________ of the box.

15. A(n) ____________________ locknut may be required if bonding jumpers are used inside of the box or enclosure.

Name: ______________________ **Date:** ______________

Social Security Number: ______________________

1. Prior to installing conductors in a long conduit run, it is a good idea to use _____ to check for hidden obstructions.
   a. a go/no-go mandrel
   b. ¼" fish tape
   c. a flashlight
   d. a blow gun system

2. If a conduit presently contains energized cables and it must be fished, you should _____.
   a. use fiberglass fish tape
   b. tape the end of the metal fish tape
   c. use a vacuum system
   d. not fish it until the cables are de-energized

3. When installing large cables which require more pulling force, select a rope that is rated for at least _______________ times the strength of the required pulling force.

4. True or False? If commercial wire lubricant is not available, any suitable petroleum-based product may be substituted.

5. When pulling vertical runs, be sure to _______________ the rope after stopping the pull to keep the cable from reversing in the raceway.

6. When installing conductors in long vertical runs, you should install _______________ to help prevent runaways.

7. True or False? When pulling long runs of cable, stopping on a regular basis will allow the cable to cool and will avoid stressing the cable.

8. True or False? Cable length meters may be used to determine the length of conductors on reels prior to making a pull.

9. When applying the capstan theory to a power puller, more wraps on the capstan will result in _______________ pulling force.

10. True or False? When pulling cables directly off of reels, you should stagger the reels to dispense alternately between the top and bottom of the reel.

notes

Name: ______________________________ **Date:** ______________

Social Security Number: ______________________

1. Cable supports should be capable of supporting ______________ times the full weight of the cable and tray.

2. Where solid bottom cable trays contain multiconductor power or lighting cables, and all cables are larger than 4/0, the sum of the diameters of all cables must not exceed ______________ of the cable tray width.

3. The two types of cable tray failure under load are ______________________

   and ______________________________________.

4. Cable tray is commonly manufactured in ______________ and ______________ lengths.

5. List the five reasons for using divider strip in cable.

   ____

   ____

   ____

   ____

   ____

6. Locate splices no more than ______________ of the span's distance from the nearest support.

7. True or False? When installing long lengths of cable, it is best to pull the cable in one continuous operation.

8. List five types of cable tray supports:

____

____

____

____

____

9. Cable tray is usually supported by _____.
   a. wall mounting
   b. trapeze mounting
   c. pipe rack mounting
   d. All of the above.

10. Cable tray is covered in _____.
   a. ***NEC Article 300***
   b. ***NEC Article 392***
   c. ***NEC Article 517***
   d. ***NEC Article 550***

11. Cable tray must be bonded in accordance with _____.
   a. ***NEC Section 250.96***
   b. ***NEC Section 300.96***
   c. ***NEC Section 300.21***
   d. ***NEC Section 392.6(J)***

12. True or False? Water piping may be run in cable tray with electrical conductors.

Cable Tray

13. When conductors 1/0 through 4/0 are run in ladder-type trays, the maximum rung spacing must not exceed _____.
    a. 9"
    b. 8"
    c. 7"
    d. 6"

14. All of the following are basic forms of cable tray *except* _____.
    a. ladder
    b. ratchet
    c. trough
    d. solid bottom

15. Cable in vertical trays should be supported _____.
    a. per the manufacturer's recommendations
    b. every 6" or 8"
    c. every 12"
    d. every 24"

notes

Name: ____________________ Date: ____________

Social Security Number: ____________________

1. Heavy-duty strippers are used to strip power cables with outside diameters ranging from ____________ to ____________.

2. Properly made splices and connections should last ____________________
____________________.

3. Aluminum has certain properties that are different than copper. These properties are:
____
____
____________________.

4. The electrolytic action between aluminum and copper can be controlled by ____________
____________________.

5. Connectors marked AL should only be used with ____________ wire, while connectors marked AL-CU can be used with either ____________ wire.

6. It is critical to strip wire properly because many experts believe that removing as little as ____________ of conductor plating from some wires can cause a failure.

7. Nearly all cable installations will require the cable to be bent at terminations and other points along the cable route (these bends must comply with the NEC®). Minimum bending radii are determined by ____________________
____________________.

8. The bending radii for various sizes of conductors entering or leaving cabinets or cutout boxes must comply with certain NEC® requirements, many of which are covered in ________
____________________.

9. All incoming lines to either incoming line lugs or main disconnects must be braced to withstand the mechanical force created by a high fault current. Detailed information on the recommended methods of short circuit bracing are normally furnished by ________________ ________________________________.

10. All wire joints not protected by some other means should be taped carefully to ____________

____

____.

Name: ______________________________ **Date:** ______________

Social Security Number: ______________________________

1. The smallest service allowed by the NEC® is _____.
   a. 208V, 30A, single-phase
   b. 240V, 60A, single-phase
   c. 230V, 100A, single-phase
   d. 120V, 15A, single-phase

2. The minimum allowed service can be found in _____.
   a. ***NEC Section 210.23***
   b. ***NEC Section 220.36***
   c. ***NEC Section 230.79***
   d. ***NEC Section 240.33***

3. One function of a capacitor in a substation is to _____.
   a. momentarily store energy
   b. protect against lightning strikes
   c. ground the system
   d. step up current

4. At the generating plant, the voltage is stepped up for transmission using _____.
   a. water turbines
   b. steam turbines
   c. air turbines
   d. transformers

5. The voltage on transmission lines is in the _____.
   a. hundreds of volts
   b. thousands of volts
   c. ten thousand volt range
   d. hundred thousand volt range

6. True or False? The conductors on the high-voltage side of a step-down transformer are called the secondary conductors.

7. What voltage would you normally find in a residence?
   a. 4,160V
   b. 120/208V
   c. 277/480V
   d. 120/240V

8. The weatherhead is another name for the ____________________.

9. True or False? Services above 400A will almost always use feed-through meters.

10. True or False? A five-dial watt-hour meter is read from left to right.

11. True or False? ***NEC Section 250.92*** is where you would look for information on bonding when using concentric and eccentric knockouts.

12. A 120/240V, single-phase, three-wire service is used to power single-phase motors up to ____________________.

13. Four-wire, wye-connected secondaries and four-wire, delta-connected secondaries are common in ____________________ and ____________________ applications.

14. The high leg is also known as the ____________________.

15. The service head should always be located _____ the service drop attachment.
   a. above
   b. below
   c. within 5' horizontally of
   d. None of the above.

16. True or False? Service switches are normally located where the service-entrance conductors enter the building.

17. Services over ____________________ normally use a group of current transformers, one for each ungrounded conductor in the service.

18. If a panel bus is braced for 20,000A, the panel ______.
    a. is capable of handling 20,000A normal current
    b. is a heavy-duty panel
    c. is rated for 20,000A fault current
    d. does not need a disconnecting means

19. The NEC® allows a maximum of ____________________ service disconnecting means per service grouped in one location.
    a. one
    b. two
    c. six
    d. no limit

20. A general purpose enclosure rated for indoor use would be a NEMA Type ______.
    a. 1
    b. 3R
    c. 5
    d. 12

notes

Name: ______________________________ **Date:** ______________

Social Security Number: ______________________

1. A standard circuit breaker contains _____.
   a. a mechanical linkage
   b. a thermal element
   c. a magnetic element
   d. All of the above.

2. All of the following are standard breaker ratings *except* _____.
   a. 10A
   b. 45A
   c. 175A
   d. 1,600A

3. The amperage interrupting capacity (AIC) is the ______________________________

   ____.

4. A current-limiting circuit breaker provides a downstream protection for ______________

   ____.

5. A ground fault circuit interrupter will trip when a fault current _____.
   a. is 1A or more
   b. is 2mA
   c. is 6mA or more
   d. reaches the rated ampacity of the circuit

6. A dual-element fuse can be applied in circuits to provide both short circuit and ______________ protection.

7. An overcurrent condition is _____.
   a. an overload current
   b. a short circuit current
   c. a harmless condition that exists on larger circuits
   d. Both a and b.

8. The point-to-point method refers to a way of calculating _____.
   a. short circuit currents
   b. the voltage drop on service conductors
   c. the fuse amp rating of a device
   d. distances between overcurrent devices

9. Coordination is the name given to the _____.
   a. sizing of all overcurrent devices from the service to branch circuit
   b. sizing of panelboards to hold the proper circuit breakers
   c. sizing of S-type fuses
   d. time-current relationship of overcurrent devices connected in series with each other

10. True or False? Edison-base fuses are outdated and should not be used in new installations.

Name: ______________________ Date: ____________

Social Security Number: ______________________

1. What does the word *normally* refer to in regard to relay contacts?
   a. The placement of the contacts when the circuit is energized
   b. The placement of the contacts when the circuit is de-energized
   c. The way the contact is most of the time
   d. The way the contact is when it has to operate

2. Because filament-type lamps have large inrush currents, standard motor control contactors must be ______________ if used to control this type of load.

3. If a solid-state relay is used to control a DC circuit, a(n)______________ is used to connect the load to the line.

4. Solid-state relays can be controlled by _____.
   a. the opto-isolation method
   b. a reed relay
   c. a latch coil
   d. Both a and b.

5. A(n) ______________ is used to provide a continuous magnetic flow to the armature when the voltage of the AC waveform is zero.

6. Mechanically-held relays and contactors use two coils and a(n) ______________ mechanism to provide for quiet operation of the circuit.

7. True or False? Most overload relays used on modern motor starters are thermally operated.

8. If the equipment fails to start with the contactor closed, probable causes would be one contact not closing, burned contacts, or a broken ______________.

9. Unlike contactors, which are generally used to switch power circuits on relatively high-current loads, relays are normally used in low-current ______________ circuits.

10. On electrical drawings, the _______________ is the symbol used for denoting low-voltage devices.
    a. square
    b. circle
    c. triangle
    d. diamond

Name: ______________________________ **Date:** ______________

Social Security Number: ______________________________

1. The limiting factor that determines the minimum time required for the nerve terminals of the eye to respond to an image on the retina is the _____.
   a. intensity threshold
   b. contrast
   c. intensity angle
   d. time threshold

2. When light strikes an object, it can be ____________________, transmitted through the object and ____________________, and/or it can be ____________________ from the object.

3. When light passes through a prism, the spectrum ranges from ____________________ on the lower end to ____________________ on the upper end.

4. True or False? The amount of light a lamp produces is directly proportional to the amount of energy the lamp consumes to produce its rated light output.

5. Which of the following types of fluorescent lamp circuits requires the use of a separately installed starter?
   a. Instant start
   b. Rapid start
   c. Preheat
   d. Bi-pin

6. The type of HID lamp that produces light with the best color rendition is the _____ lamp.
   a. metal halide
   b. mercury vapor
   c. high-pressure sodium
   d. low-pressure sodium

7. True or False? When installing a bi-pin fluorescent lamp, it should be rotated ¼ turn in its lampholder to seat and connect it.

8. Fluorescent troffers are a type of _____ lighting fixture.
   a. surface-mounted
   b. recessed
   c. suspended
   d. track-mounted

9. When opening a shipping carton containing a light fixture, make sure to open it __________ so as not to ______________ the fixture.

10. True or False? When installing a surface-mounted fixture that weighs more than 50 pounds, it must always be supported independently of the outlet box.

11. Recessed fixtures not labeled Type IC must be installed so that they are at a minimum distance of ____________________ from any adjacent insulation.

12. A fixture whip connected between a junction box mounted above the suspended ceiling and a fixture installed in a suspended ceiling must have a minimum length of __________________ and be no longer than ____________________.

13. When fluorescent troffers and other fixtures are installed in a suspended ceiling, the ceiling framing members must be __________________ __________________ together and to the ____________________ structure.

14. True or False? Lighting fixtures and/or accessories for track lighting systems can be used interchangeably between systems made by different manufacturers.

15. When cutting the branch circuit wires in preparation for connecting a light fixture, you should leave at least _____ of free conductor length at the outlet box where the wire emerges from its cable sheath or raceway.
   a. 3"
   b. 5"
   c. 6"
   d. 8"

CONTREN® LEARNING SERIES — USER UPDATE

NCCER makes every effort to keep these textbooks up-to-date and free of technical errors. We appreciate your help in this process. If you have an idea for improving this textbook, or if you find an error, a typographical mistake, or an inaccuracy in NCCER's Contren® textbooks, please write us, using this form or a photocopy. Be sure to include the exact module number, page number, a detailed description, and the correction, if applicable. Your input will be brought to the attention of the Technical Review Committee. Thank you for your assistance.

Instructors – If you found that additional materials were necessary in order to teach this module effectively, please let us know so that we may include them in the Equipment/Materials list in the Annotated Instructor's Guide.

Write: Product Development and Revision
National Center for Construction Education and Research
P.O. Box 141104, Gainesville, FL 32614-1104

Fax: 352-334-0932

E-mail: curriculum@nccer.org

Craft ______________________ Module Name ______________________

Copyright Date ____________ Module Number ____________ Page Number(s) ____________

Description

____

____

____

(Optional) Correction

____

____

(Optional) Your Name and Address

____

____